CALEB ABBAS

OFF GRID SOLAR

The Ultimate Guide to Solar and Other Green Energy Sources, Discover All About Green Energy and How You Can Use Them to Help Save the Planet

Descrierea CIP a Bibliotecii Naţionale a României
CALEB ABBAS
   OFF GRID SOLAR. The Ultimate Guide to Solar and Other Green Energy Sources, Discover All About Green Energy and How You Can Use Them to Help Save the Planet / Caleb Abbas – Bucharest: Editura My Ebook, 2021
   ISBN

CALEB ABBAS

OFF GRID SOLAR

The Ultimate Guide to Solar and Other Green Energy Sources, Discover All About Green Energy and How You Can Use Them to Help Save the Planet

My Ebook Publishing House
Bucharest, 2021

TABLE OF CONTENTS

Foreword

Last 2006, the world was rocked by the documentary presented by former United States Vice President, Al Gore.

An inconvenient truth was a documentary about how the earth's temperatures are rising, the polar ice caps are melting and the oceans are rising. Decade after decade we hear news about climate change, increasing pollution with more and more animals and people getting sick. Basically, the story was all about how the earth will die if we won't do something about it. Get all the info you need here.

Going Green Energy

Learn About Energy Sources That Will Help Save The Planet

CHAPTER 1

INTRODUCTION

Synopsis

Global warming and pollution go hand in hand but the major concern of most scientists and environmentalists now is global warming or the constant rise of the earth's temperature.

For the past decades, science has recorded that the constant rise in temperature has been caused by human activity. With all the pollutants that are produced by humans, these three gases namely, carbon dioxide, nitrous oxide and methane.

Studies have already shown how these gasses can trap heat but there are still some uncertainties on how the earth reacts to it. What they do know is that the earth needs to stay temperate enough to support life but that humans are speeding up the process.

The Basics

With the rise of our technical advancement, humans have produced more and more greenhouse gasses while the earth is struggling with constant climate change. As humans, we create more and more factories to provide everyone with all the materials and appliances that they need, more pollutants are dumped into our air and water supplies.

During the early years, factories or companies would improperly dump both chemical and organic waste into the earth. Then, as time went by people suddenly noticed the changes and the effects that their actions did. Most of the people were concerned but there were also those who couldn't care less if the earth was already sick.

If we are not careful, we could end up rendering the earth unlivable or we will cease to exist due to natural calamities. The constant rise in temperature is also affecting our weather that is why we experience out of this world seasons. Farmers and consumers alike will also have to be concerned about what is happening because crop seasons will also change with the season.

The story became a hit and everyone was either screaming foul about it or, praising the creators for opening the eyes of all the people.

Everyone has been abusing the earth and neglecting all the signs and effects that this has caused. Now, people are more open and understanding, almost everyone is now willing to share and help save our mother earth.

Recently, the green revolution has started to pick up in the communities. People are now looking into alternative energy resources, recycling, tree planting basically anything or everything that can help reduce the effects of our negative impact on earth.

People are now starting to be conscious on what they are doing and using in their everyday lives. From homes to offices even up to the cars that people are driving, the going green revolution is starting to catch on.

In the homes, people are encouraged to use energy saving bulbs, control the temperature of their water and even their oven. Simple tips like this will greatly help the earth recover and it will also give everyone a chance to help out even in the simplest ways.

CHAPTER 2

WHAT DOES GREEN ENERGY MEAN

Synopsis

Green energy is the name that is used to supply, call or classify all the different forms of resources that are eco-friendly and non-polluting. The most common examples are type of power resources the humans have found throughout the years. These samples are the geothermal, win, solar and hydro resources.

A controversial member of this group though is nuclear energy. This power supply already has the two basic requirements to be considered as a green energy source, less pollution and low carbon emissions.

Due to all the controversy and dangers that it possesses though, most of the people do not consider this as a clean source of energy. Nuclear power plants are filled with radioactive materials; waste and their reactors can create or end up in horrible meltdowns.

What Does It Mean

The most common form of green energy though is the production of energy resources for the whole world. The United Nations (UN) is leading the campaign all throughout the different countries that promotes the use, creation or discovery of green energy. They are urging all the different institutions, businesses, offices or organizations, to switch to more energy saving materials and energy resources and help save mother earth.

They are also promoting the purchase of green power to help fund the research and development programs for renewable energy and, aid in clean-up drives or programs that can help reduce or clean-up all the waste the humans have left with mother earth.

The UN also wants people to know that with renewable energy we can actually have a perpetual supply of power with less or no pollution at all. In support of their cause for a greener earth, UN also distributes green tags of certificates. These will allow various consumers and businesses to be updated on the campaign and show their support to the cause.

Green energy can be associated with renewable sources of energy, alternative solutions, Eco friendly technologies, non-polluting energy and even the source of fossil fuels. Green energy basically stands for energy that was developed from green sources. In contrast to this though, Brown energy is the name given to resources that do not use earth friendly materials or methods.

There is no energy source that has a zero production of pollutants but with the going green suppliers, pollution is kept close to a bare minimum. Resources like the geothermal energy, solar energy, tidal power, wave power fall, small scale hydro power, anaerobic digestion and more are renewable resources of energy that can used for everyday consumption.

In terms of energy efficiency though, people say that renewable resources produce a significantly lower level of power, which is why we have "wind farms" or solar farms. Others say that if you want to create a renewable resource even for your home, you should be prepared to shell out some serious cash. Both of them do have points but what matters is that it is doable and that they are doing the earth a huge favor. Plus, if you want to save up from your energy bill and win with more rebate points, you should inquire about it at your local government units. Depending on different states, they have 2-3 programs that will help you set your equipment and reimburse you for all your hard work.

CHAPTER 3

ALTERNATIVE FUELS

Synopsis

With the spark of the green energy in the scientific world, people have gone on to discover alternative fuels. These are the types of energy resources that can replace their more expensive versions like diesel and gasoline. Plus, they are very eco-friendly that more people have taken an interest in them. With their discovery, we can already help the conservation of natural resources that are needed to produce gasoline.

Fuel

With all our advancement in science and technology, alternative fuels are now readily available to anyone who may want to use it. The most common use for alternative fuels is powering up our cars. Ever since its discovery, numerous cars are already being powered with liquefied petroleum gas, methanol, ethanol, electricity and even compressed natural gas. The best part about this change is that alternative fuels are way cheaper than their oil based counterparts and you can also travel without sacrificing the mileage of your car. Without the harmful substances or residue that comes with gasoline, alternative fuels are proven to be very clean and will clearly help reduce air pollution.

Alternative fuels can help save the world and pollution can be avoided if we just simply try them out. A basic example is an electric powered car. Without all the harmful chemicals that gasoline produces, there will be lesser chances that air pollution will worsen for every mile that you travel on wheels.

To give you a feel on what our earth eats up while you drive, here are the harmful chemicals or pollutants that are emitted by your car for every mile that you cover - Carbon dioxide, carbon monoxide, nitrogen oxides, particulate materials and unburned hydrocarbons. Imagine all of that and multiply it by every single car in the whole world. Can you imagine it now?

With alternative fuels, there will be none of that. With their use, people will be able to minimize all of these pollutants or get rid of all of them completely.

Another good thing about alternative fuels is that its components can be found everywhere, even in your own backyard! Imagine the following items, recycled paper, plant waste, animal feces, used vegetable oil, or old frying oil as power for your vehicles. Isn't it amazing? Don't think about stuffing all of this into your gas tank though, everything still has to undergo different procedure to make it useable for your vehicles.

In line with the discovery of alternative fuels, people have also developed specific vehicles that can only run with the designated fuel made for it. The best example for this would be the electric cars. Electric cars no longer need regular fuels or even the alternative choices. Electric cars run on electricity and can be recharged via an outlet. It's just like charging your cellular phones.

Unfortunately, though, not all good things are perfect. Alternative fuels may be the ideal solution in lessening air pollution but it is also expensive since all the materials have to undergo specific processes. Also, gasoline or diesel can also provide better mileage for your cars since it produces more power. Alternative fuels cannot do that just yet so people will just have to settle with cruising down the street and not zooming by.

CHAPTER 4

FUEL CONSERVATION TECHNIQUES

Synopsis

Fuel is one of the most expensive and sought after commodity in our world. We use it for our transportation, for other businesses and for manufacturing companies to name a few. Due to it's never ending price increase, every person who drives or owns a car need to learn how to conserve his or fuel so they can go on the best road trip ever!

Conservation

Properly inflate your tires

Drivers must always remember to check their tires. This is probably one of the cheapest and easiest way to control a person's fuel intake but it is often overlooked. If a diver will travel with tires that lack air, they will end up burning more fuel to gain more power to move on.

Weight Check

Make it a practice to clean up or de-clutter your car. When they are loaded with too many heavy things, the tendency of most machines will be to bring out more power thus using up more fuel. According to studies, for every 200lbs of weight added to your car, it will trim about one mile off your fuel efficiency.

Buy in Morning

It is probably not a known fact that consumers are charged based on the volume and not the density of what they are buying. This is why they say its best if you purchase gasoline early in the morning when the temperature is cold enough and gasoline will be at its densest.

Idling

Did you know that when your engine is idling for a longer period of time, you are probably consuming gallons of gas but your mileage is zero? That's probably one of the saddest things ever that you will see on your dashboard! Plus when you are restarting your engine, you are more or less using up the same amount of gasoline when you are idling by for about thirty seconds. So remember, when you are pulling off the road to make a call or text to someone, just do yourself a favor and turn off your engine.

Scheduled Maintenance

Just like any appliance, cars must also be well maintained. If you properly maintain your vehicle's air filters, you will have better fuel economy. Remember that when your engine's air supply is clogged up, you will end up with a higher air to fuel ratio. Whenever this happens, your engine will try and make up for it and end up using up more gasoline.

Energy Conscious

Remember those tips about turning off lights and unplugging etc? Well, it also applies to your car. Drivers should always practice energy saving with their vehicles as well. The best example of this would be to turn off your air conditioning if you are not using it. It can reduce 5% to 20% of your gas consumption and drain your engine power if you're not careful.

Follow the Speed Limit

If a driver adheres to the speed limit and anticipates the traffic flow, he will be able to save up on his fuel. If you keep on traveling at fast speeds you will only end up burning up more fuel. You're more likely to spend around 15% more of your fuel if you drive at 65mph rather than 55mph.

Plus, if you already anticipate the traffic flow, stops and other road changes, you won't have to subject your vehicle to jackrabbit stops or starts and consume more fuel.

A driver can probably save about 20% of his fuel economy if he chooses to drive smoothly rather than aggressively.

CHAPTER 5

SOLAR POWER

Synopsis

Ever since the world began, the sun has already been in existence. It was already powering up the earth, providing energy to the various life that spawned there. It became the lifeline of animals, humans and even the microbial members.

Plants, with the help of their photosynthesis can create almost all the necessary food that they need. Before manufacturers or scientists discovered the power of the sun, people relied heavily on fossil fuels and didn't think much of the warmth they got from the sun.

Lately with all the advancements in technology, scientists and developers have finally introduced to the general public the use of solar panels.

Solar Power

There are two ways in which we can benefit from the Sun's energy. One is through photovoltaic (PV) devices or what we call, solar cells. PVs harvest the sun's energy by changing the sun's light directly into electricity through the cells. It is often used in areas that cannot be reached or are not connected to the electric grid. You can also find PV cells being used to power calculators, watches or even road signs that come with lights.

The other one is through, Solar Power Plants that generate electricity when the harvested heat from their solar thermal collector heats the fluid that produces steam and is used to power the generator. During the last survey in 2006, there were already 15 solar power plants being operated in the United States. 5 are in Arizona and 10 are in California. There may be smaller and undocumented plants though since those that are producing less than 1 megawatt are not included in the statistics.

Photovoltaic energy is what you call the conversion of the sun's light into electricity. The Photovoltaic cell or its common name, the solar cell, is the technology that was developed to help people convert solar energy useable electricity. It is non-mechanical and is made up of silicon alloys.

So, how does Photovoltaic cells work you ask? Well, it's all in the Photons (the sun's energy) and electrons (energy). As the energy from the sun's photons hit the cells, some are absorbed, ignored or just pass through. The absorbed photons are the ones that knock out the electrons and produce an imbalance. The imbalance that was produced by the process is what you will be harvesting. Then as electrons start to move, the battery hooked up to your panel will then serve as their path and turns into electricity.

Solar thermal power plants use the sun's heat to produce enough energy to turn the fluid into steam. The steam is what will make the turbines move and start to power the generator to produce energy. Fundamentally, the power plants work the same way as plants that use fossil fuels. The difference lies in what power they use to move the turbines and produce electricity.

The beauty of solar energy lies in the simplicity of its concept, harvesting the heat of the sun. It is one of the most practical of all renewable resources and the easiest. People may be discouraged with the price to set up their own solar panels but remember, you can research if your state offers rebates or help when you decide to convert your home into solar power. Solar energy is something that you should spend more on for you to finally save more for the future.

CHAPTER 6

WIND POWER

Synopsis

Right next to the sun, the wind is one of the most accessible renewable resources that we have in our hands.

The sun hits the earth's surface at different intensities all throughout the day. Wind is produced when hot air rises up and encounters different temperatures from different areas. As the wind is produced, windmills catch this energy and use it to power their turbines and then create electricity.

Wind energy has also been around for quite some time now. Unlike solar power though, wind does not need sophisticated technology to be able to produce energy. Windmills have been made since during the olden times to help produce enough energy or power to pump water or grind the harvested grains.

Using Wind

It also has become one of the fastest growing renewable resources in the whole world. In 2005 alone, the technology has seen so many improvements that in the United States and Europe, the generating capacity of the wind-power increased by 27 percent and 18 percent respectively.

One of the major advantages in using wind energy is that it is the cleanest source of electricity. The turbines only "harvest" the wind and no other chemicals or equipment is needed, to power their generators. Windmills are also very cost-effective since they are low maintenance and expenses for operating them are minimal.

Windmills are clean resources for energy for they do not produce any air or water pollution. No fuel is burned or used in any other way to power the turbines so there is no toxic waste. Unlike other power plants, who are still using up carbonaceous fuels and who are producing a lot of waste that needs to be disposed of properly.

Another advantage of the Windmills is that despite of being low maintenance; they can still generate or increase 27% more jobs than those who produce energy from other resources.

Not all things are immune to shortcomings though, including the clean and renewable resource of wind. Wind energy experiences intermittent production since wind is not constant. They can only produce enough energy when the velocity of the wind is at a certain level. When the wind is too strong, the turbines need to be stopped and wrapped up to avoid incurring any damage. When the wind is too slow though, the turbines won't be able to move enough and produce energy. Due to this shortcoming, there will always be a need for backup stations that will do all the work during windless day.

Due to this fact also, wind energy cannot be used or relied upon as a major contributor to the national power grid. It will be hard to balance out the power distribution when your resource is never fixed.

Other challenge that windmills face is their effects in the common household. Most of the time, localities complain about the turbine noise or the potential effects it will have on the wildlife for it will be a common hazard to flying birds.

Wind energy may not be a consistent supplier of energy but it is already a proven and tested clean source of energy. With the proper settings and preparations, you may even install one at home provided that you take into consideration the common complaints or the concerns of your neighbors.

CHAPTER 7

BIOENERGY

Synopsis

Another great source for renewable energy is bioenergy. This source is created or made available though different biological resources like biomass. Any material that has sunlight stored inside them in the form of chemical energy is called a biomass.

Bioenergy is the source of energy that can be obtained from biomass. It is said that biomass is the fuel that's being used while bioenergy is the energy that can be extracted from that fuel.

Bio

Biomass is a type of renewable source that is from biological materials that are obtained from the recently living or living organisms like animal based materials, vegetables and plants. Other materials like animal manure, slurries, straws, husks, poultry litter, trees and agriculture can also be used to produce biomass.

In simpler terms, bioenergy is the fuel that is provided by recently living materials that are produced naturally such as animal fat, wood or plants. In a more complicated or scientific explanation, utilizing bacteria that are genetically modified so that it will create cellulosic ethanol can also produce Bioenergy. Oil and coal are also organic matter but they are not considered as sources for bioenergy since they were not living things in the first place.

One of the major impact that bioenergy will have in our lives is that is can replace our existing diesel and petrol with little or no changes or modifications to the engines. Bioethanol and biodiesel are already making their names known that even some notable vehicle manufacturers are already pledging to support this discovery. Plus, with the go green movement, news about biodiesel and its positive impact on our environment have become more widespread.

There are different types of bioenergy and they can be produced through different ways and they also have different uses.

By fermenting the starch or sugar portions of certain agricultural materials, bioethanol can be produced. The most common materials used for this are sugar cane, sugar beet or maize. As of today, the largest producers of bioethanol are the United States, China and Brazil. It has also been reported that majority of Brazil's vehicles are already powered by bioethanol.

Biodiesel is extracted from both inedible or edible plant oils like palm oil or grape seed. It has also been noted that waste-cooking oil is also useable for biodiesel.

Carbon dioxide and methane are used to create what we call biogas. It is produced from biomass such as biodegradable waste, sludge, sewage, feedstock and even manure. It can be used as an addition to natural gas to be used in vehicles or burned to produce energy.

The solid types of bioenergy are wood, charcoal and biomass pellets. These items are basically burned to heat or to produce electricity.

Promoting or using bioenergy is greatly beneficial to the earth not only because it can potentially replace fossil fuels, but also because it can help lessen our waste. It is the proof that there is indeed money or more uses for what we see as trash and that if we use them to the best of our abilities, we will be able to use them to produce a cleaner and more sustainable form of energy.

CHAPTER 8

ENERGY CONSERVATION TIPS

Synopsis

Ever since news about global warming, pollution and the slow decline of our energy supplies, people have become more conscious of how they can help conserve energy. People are now finally becoming more aware of the situation of our environment and they want to do something about it so future generations will still have a world to live in.

All generations, from adults to children, are now learning and relearning thing that they can do to help save mother earth. Almost every home now is more conscious of energy conservation tips not only for their budget, but also for the benefit of their surroundings.

Conserving Energy

People may say that they are too busy with their lives, work or school that they do not have time to worry about more things. This should not be the case because even the little things that are done are important and useful for a greener nation.

Here are some simple tips that you can do in a matter of minutes and can even be part of your daily routine.

Remember to turn off your lights.

When you're done with your work or if you're going out of the room, remember to turn off the lights that you are no longer using. This simple action will not only help you save your environment, it will also save your bill!

So remember,

If you're on your way out, turn off the lights. If you're done in the kitchen and you're on your way to the bathroom, turn off the lights.

Bottom line is, if you're not using it anymore turn your lights off!

Unplug

Another energy cruncher is leaving chargers or appliances plugged in even when you're done using them. What you might not know is that even if they are turned off or not connected to anything, they are still eating up energy.

Even the tiny LCD screens on your home theater or the standby lights on your sleeping computer will still contribute to your electricity bill.

Sleep and Hibernate

Admit it, you love to work on your computer and sometimes you just have to leave it on to finish some downloads or to avoid reloading everything. If you just can't bear to shut down, choose to set your computer to sleep or hibernate. That way, at least your energy consumption will be lesser since your monitor will be turned off.

Check your temperature

A lot of people say that the lower your temperature for your heater, the less energy it will eat up. This is indeed true but. If you don't regulate it properly you'll end up running out of hot water and using up more energy to reheat more water for the shower or dishwasher. Look for a steady temperature and stick with it.

You should also remember that preheating your oven for a longer time than needed would only eat up more energy and not improve your dishes. Plus, your refrigerator doesn't need to freeze everything inside.

Remember, balance is the key.

CHAPTER 9

GOING OFF THE GRID

Synopsis

Now that you've read all about the different types of natural and renewable energy, I'm sure you're thinking of the possibility of going off the grid. The idea of living in a self-sustaining home is so appealing not only to your pockets but also for the great good of the environment. But, before you do this you have to check make sure you are already ready and prepared to go off on your own. You will no longer have a help line and you'll also be producing enough energy to supply all of your needs.

Do not be misinformed too, going off the grid does not mean that you will be choosing a power source like wind or solar then have the power companies hook you to it. That is not considered going off the grid. If you truly want to go off the grid, you will have to produce your own energy and not rely on outside companies to help you out or deliver their services to you.

Grid Free

You might think it will be hard but don't fret! Houses that are off the grid are not new and have been around for a while. Just imagine those houses in remote villages that can't be reached by service providers. Due to necessity, they figured out a long time ago how to survive on sustainable energy resources like wind and solar power. So, don't be afraid to take that step. Depending on your knowledge, experience and interest you can choose to slowly go off the grid or you can just sustain a small part of your house.

In some cases though, some families can produce enough electricity to sustain their homes and still have extra power. In some cases and depending on your area, the residents can actually sell their excess power back to the power companies and make money. For years, a lot of homes have been doing this with positive results.

Now that you have an idea on what it truly means to go off the grid, the next step is to choose the sustainable power source that suits your needs and budget. Do your research properly and intensively so that you can make a proper informed decision. The most popular options are solar and wind power. For both, you need money and time plus you'll have to modify your house so it can adapt to the change in power supply.

Aside from helping the environment, going off the grid can give you a certain confidence that most power companies will not. You are no longer under their influence plus in case of emergencies like storms or during a power outage, your home will not be part of their affected lines and you'll continue on with your self produced heat and electricity.

It may seem like a crazy and overly expensive idea at first glance but rest assured that with a little more research and inquiry, you wouldn't have to spend as much what other people will think. With the rising interest of people in self-sustaining energy providers, solar and wind technologies will eventually become mainstream and then prices might fall.

You are encouraged to think out of the box if you want to go off the grid. You have to plan and to not lose faith in it.

CHAPTER 10

FINDING PLANS AND KITS FOR GREEN ENERGY AND THE BENEFITS

Synopsis

If you have decided that you want to try going off the grid, the next step would be to choose the right plans and kits that will fit your budget and your needs. So far, you will find numerous providers of solar and wind energy kits that you really won't have any trouble, unless of course if you think your budget will not be enough or if you want to save up on construction costs. If that is the case, one option that you could follow is to make your own solar or wind power systems.

Here are some tips and a simple outline for you to follow if you wish to start on your do-it-yourself wind and solar power systems.

Tips

Planning

Now that you have all the data and plans in your papers and head, you can go ahead and purchase your DIY solar and wind power systems kit. There a lot of them available online and they also come with comprehensive instructions.

All you have to do is go over them and choose which one is more energy efficient and which one you like best. Be wary of some kits that have subpar qualities for you might end up spending more on repairs if you go for them. Make sure that what you're purchasing has clear instructions and is proven effective.

Also, make sure that you have already mapped out your electrical wiring or whatever needs to be done before setting up.

Building

If you already have your plans on hand, it's time to build your solar/wind system. Carefully read all the instructions and make sure that everything is in top quality. You don't need any experts as long as you know how to read instructions and then you'll be successful in building your own panels or turbines.

You also have to make sure that your area and your home is ready to accept the new changes. You have to check with your local neighborhood if you're allowed to put up a windmill or if your solar panels will bother them at a certain area. These may seem quite trivial at first but after a few researches, it seems that the best option would be to check with them. It is always better to make informed decision rather than having to remove all your hard work from your lawn since the neighbors are complaining about it.

Wrapping Up

Building a new energy system will not be easy. You will be building your own power system and not just a mock toy. You have to analyze and make sure that everything is done right so you won't waste your time or your budget. With patience, everything is possible! You'll never notice but eventually, you'll be done in no time!

Building your own solar energy and wind energy systems nowadays is highly beneficial for you since ordering them is quite pricey. You will not only help save the environment but you will also help your family and your budget. It will be the start of something new for your family with a healthier life style.

Printed by Libri Plureos GmbH in Hamburg, Germany